6th Grade Math
Volume 7

© 2013 OnBoard Academics, Inc
Newburyport, MA 01950
800-596-3175
www.onboardacademics.com

Table of Contents

Measures of Central Tendency

Key Vocabulary

average

mean

mode

median

range

Share the money evenly.

You can create an equation to share the money evenly.

$$4 + 2 + 3 + 7 = 16$$

$$16 \div 4 = 4$$

Each person receives $4.

The Mean

On the previous page, we calculated the mean amount of money.

To calculate the mean:

Step 1   Sum *the set of terms*: in this case, the money.

Step 2   Divide the sum by the number of terms: in this case, the number of people who will share the money.

Find the mean value: {3, 12, 15}

Step 1:   =

Step 2:   =

What is the mean number of cookies on a plate?

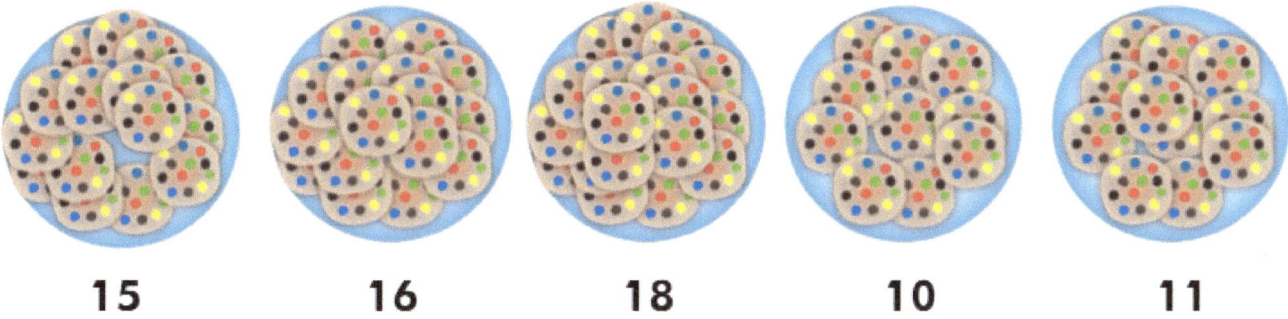

15   16   18   10   11

What is the mean height of the students?

Practice calculating the mean.

72"   68"   64"   68"   66"   61"

What is the mean height in feet and inches?

What is the mean height of the boys?

What is the mean height of the girls?

A new student joins the group.

$$64 + 68 + 72 + 61 + 66 + 68$$

A new student joins the group and the mean height increases from 66.5 inches to 67 inches.

How tall is the new student?

Name_____

Measures of Central Tendency Quiz

1 True or false, the mean of {845, 866, 888, 987, 999} is greater than 888?

2 What is the mean of {100, 125, 150, 175, 200}?

- **A** 100
- **B** 150
- **C** 125
- **D** 125.5

3 The mean of {6, 2, 9, n} is 7.5. Find the value of n.

4 Find the value of n if the mean of {6, 2, 9, n} is 7.5

Line Plots

Key Vocabulary

line plots

Company policy is 12 nuggets to a box.

The McCluck Nugget Company received complaints from several customers who said that when they purchased McCluck Nuggets, there were never the same number of Nuggets in a box.

The company decided to randomly survey some of their restaurants to see if they were following company policy of 12 nuggets per box.

A single purchase of Nuggets was made at an unspecified number of restaurants and the results were organized in a line plot.

No. Nuggets in a box	No. Boxes
9	3
10	5
11	7
12	5
13	0
14	1

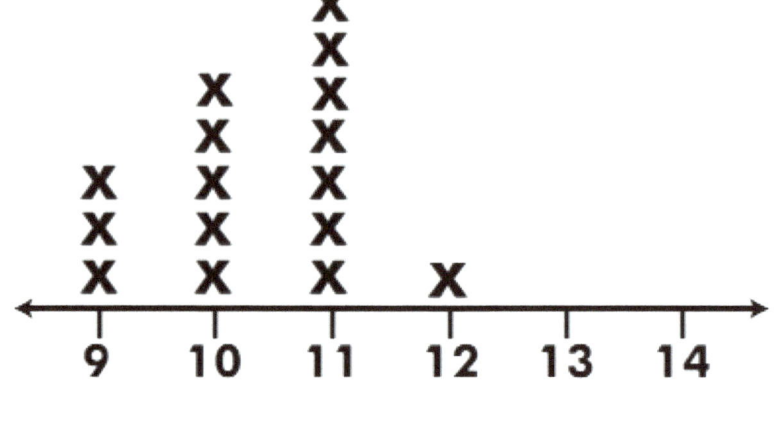

The Data

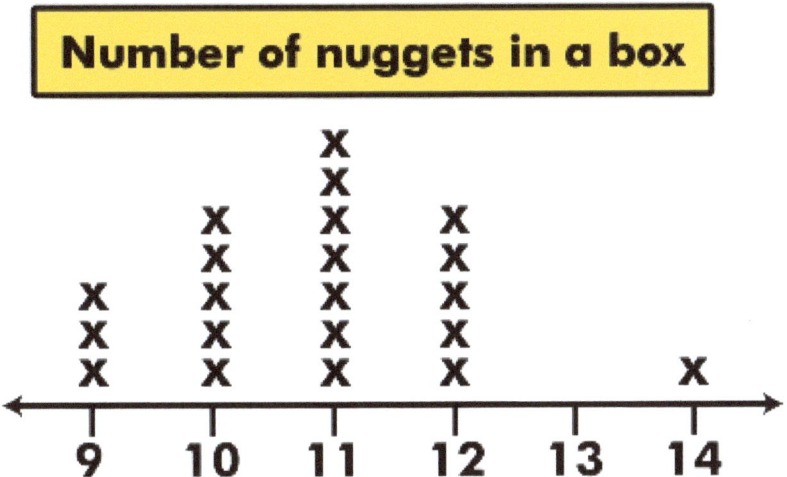

Describe the distribution of the data.

Are there any gaps?

Are there any clusters?

Find the mode.

Mode-The number that appears the most often in a set of numbers.

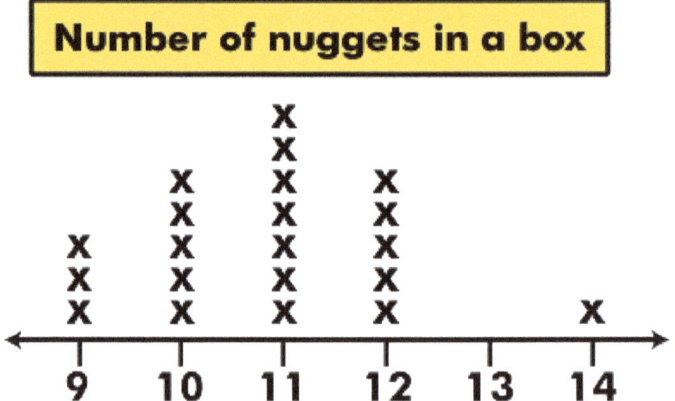

How many boxes of nuggets were purchased?

What is the most common number of nuggets in a box?

What is the mode?

Find the mean.

Mean-average of the numbers, the calculated central value of a set of numbers.

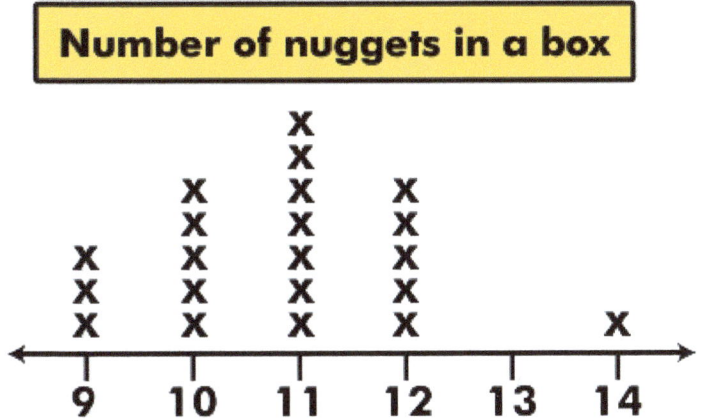

What is the mean* number of nuggets in a box?

Were customers right to complain?

*Rounded to nearest tenth

Find the median.

Median-the middle number in a sorted list of numbers.

How many boxes of nuggets are represented on the plot?

What is the median value?

Find the range.

Range-the difference between the lowest and highest values.

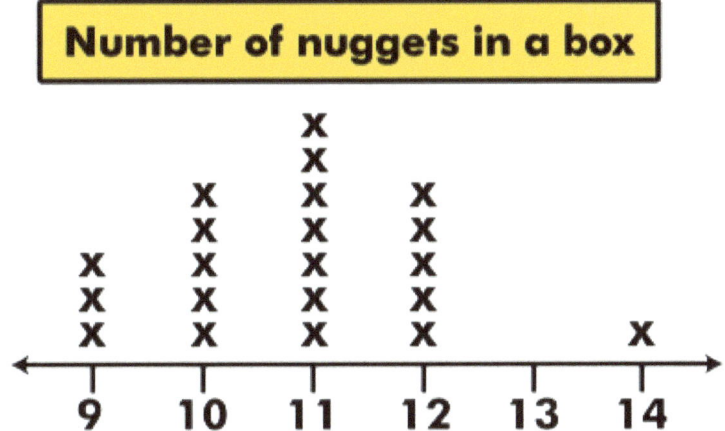

What is the greatest number of nuggets in a box?

What is the least number of nuggets in a box?

What is the range?

Memo from head office and you're the boss.

Write a memo from Head Office to all McCluck outlet managers describing the results of the survey, and reference measures of central tendency in your memo.

_____

_____

_____

_____

_____

_____

_____

_____

_____

_____

_____

_____

_____

Name_____

Line Plots Quiz

1 True or false? A line plot is useful for displaying large data sets with a large range.

2 Find the mode score for the science test.

3 Find the median score for the science test.

4 Find the range for the science test.

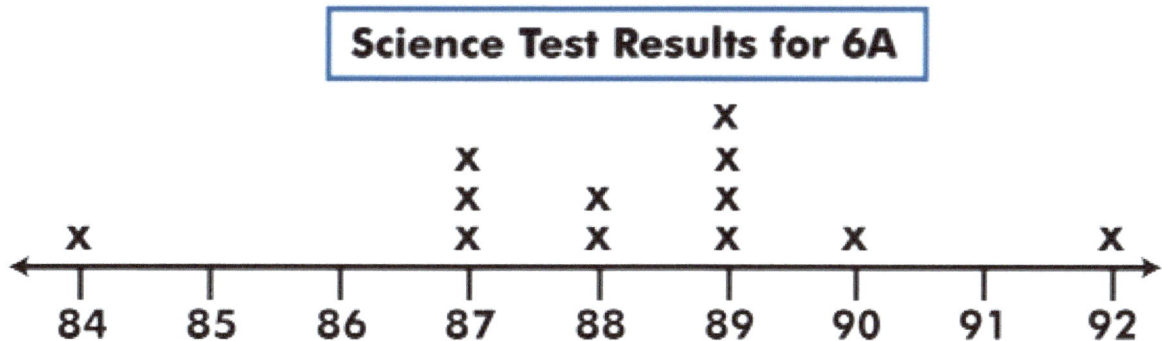

Science Test Results for 6A

Circle Graphs

Key Vocabulary

circle graph

percent

Class Survey of Pet Ownership

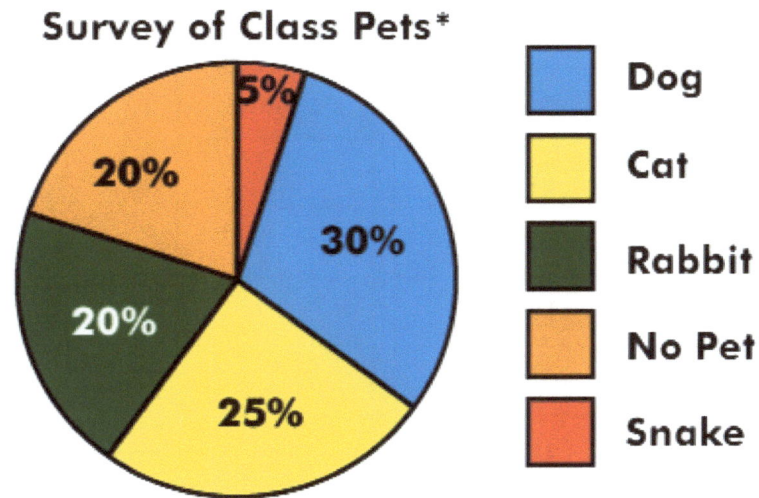

There are 20 students in the class. How many own a dog?

How many own a pet with four legs?

*Assume all students own only one pet and no pets have lost legs!

Drawing a Circle Graph.

Complete the table below to draw a circle graph.

Honest Auto – Q1 Sales

Type	Percent	Degrees
Trucks	20%	°
SUVs	30%	°
Hybrids	10%	°
Sedans	40%	°
TOTAL	100%	360°

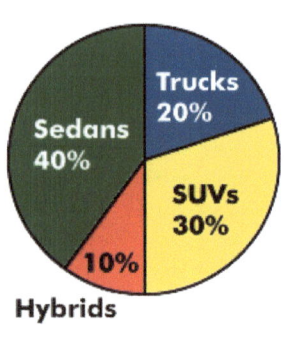

Votes for Class President
Draw a circle graph using the information below.

Election Results

Student	Vote
Michael	25%
Carmen	20%
Ashima	40%
KJ	15%

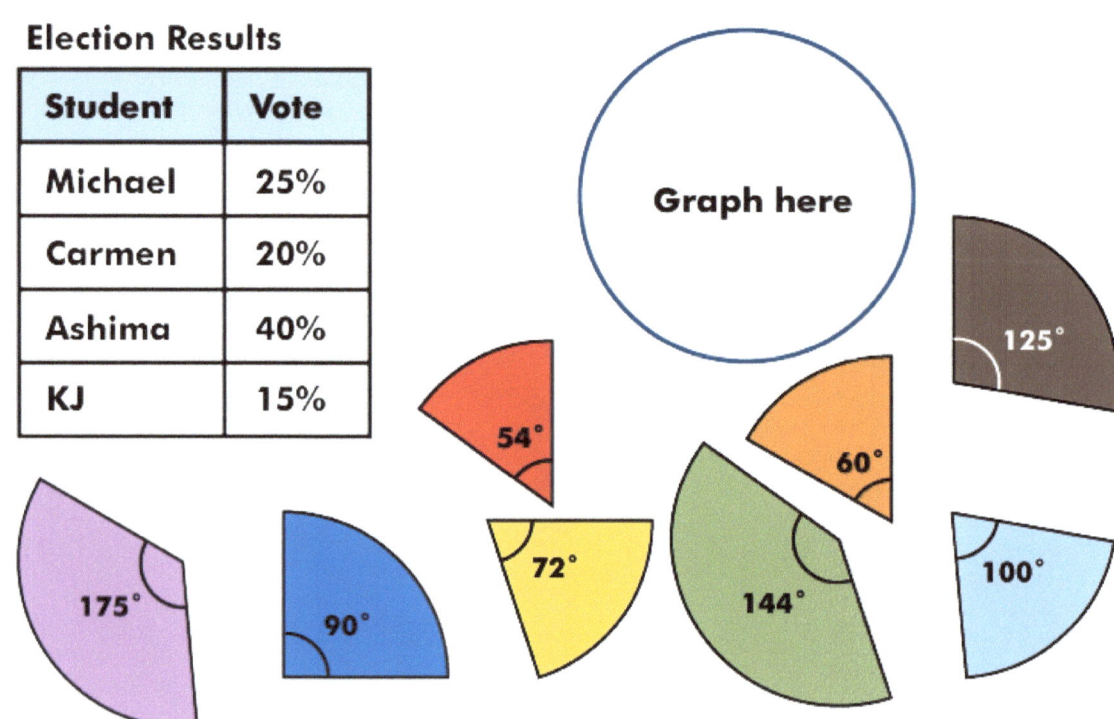

Interpreting Data

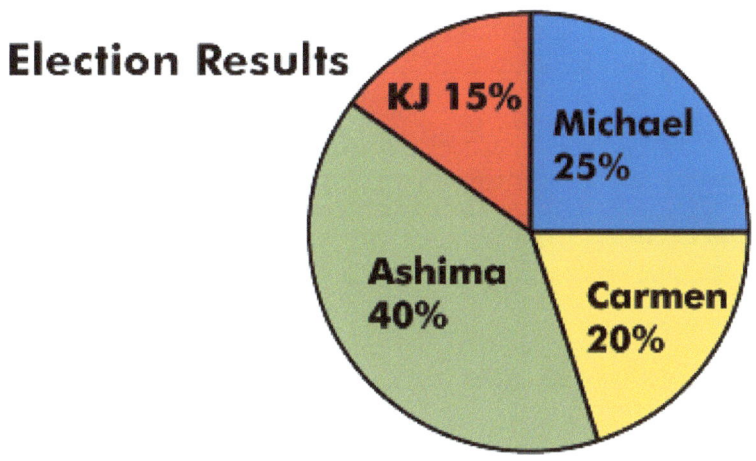

If Ashima received 16 votes, how many students voted?

How many votes did Carmen receive?

How many votes did Ashima win by?

Blood Type in the USA

Blood Type	% of US Population	Degrees
O	45%	°
A	40%	°
B	11%	°
AB	4%	°

There are about 300 million people in the US.

How many people have blood type O?

Which circle graph matches the data?

Alicia's vacation earnings

🟥	Dog walking	$624
🟩	Babysitting	$600
🟧	Pizza delivery	$864
🟦	Birthday money	$312

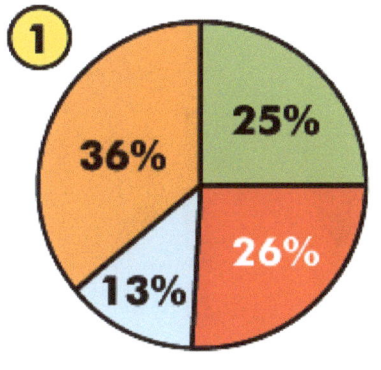

1
25%
36%
26%
13%

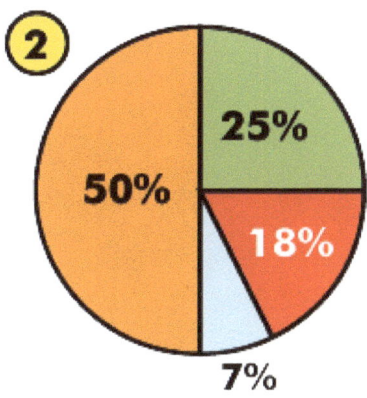

2
25%
50%
18%
7%

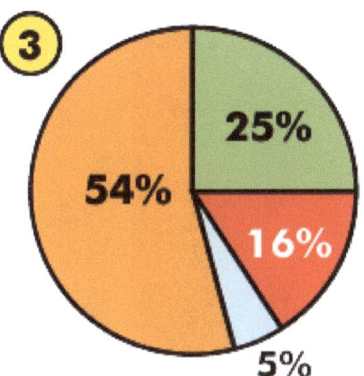

3
25%
54%
16%
5%

Name_____

Circle Graphs Quiz

1 True or false? There are 270° in a circle graph.

2 There are 240 students in Grade 6. How many students said apple is their favorite fruit?

A 60

B 72

C 84

D 80

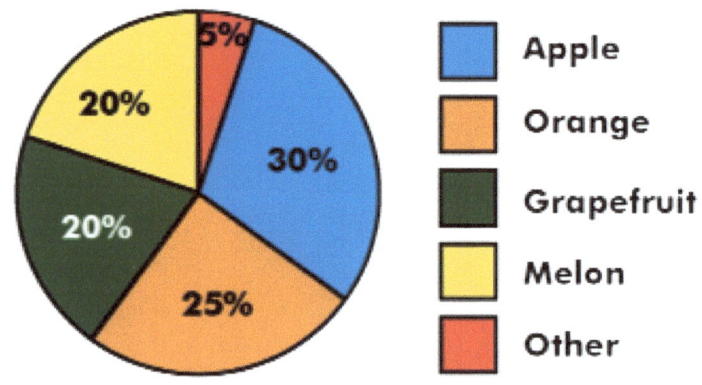

Favorite Fruit G6 Students

Apple
Orange
Grapefruit
Melon
Other

3 How many students prefer oranges?

4 How many degrees are represented by "other" fruits?

Probability

Key Vocabulary

probability

event

outcome

tree diagram

sample space

Use the number for each statement and place it on the probability number line.

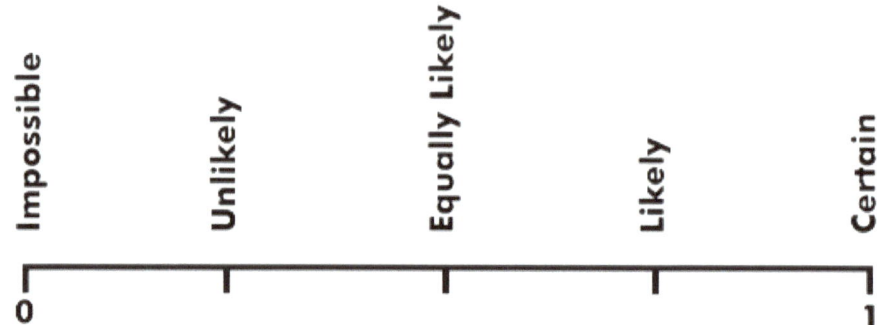

1 I will use a computer some time tomorrow.

2 I will flip a coin and get a head.

3 Abraham Lincoln will serve three terms as President.

4 I will flip a coin and get a head or a tail.

5 The Principal will beat Tiger Woods at golf.

Finding Probability

Color	Number
Red	
Green	
Blue	
TOTAL	

Probabilities can be written as fractions, percents or decimals.

If pick out one ball at random, what's the probability the ball is:

● Green $\dfrac{3}{10}$

● Red

● Black

Finding Probability

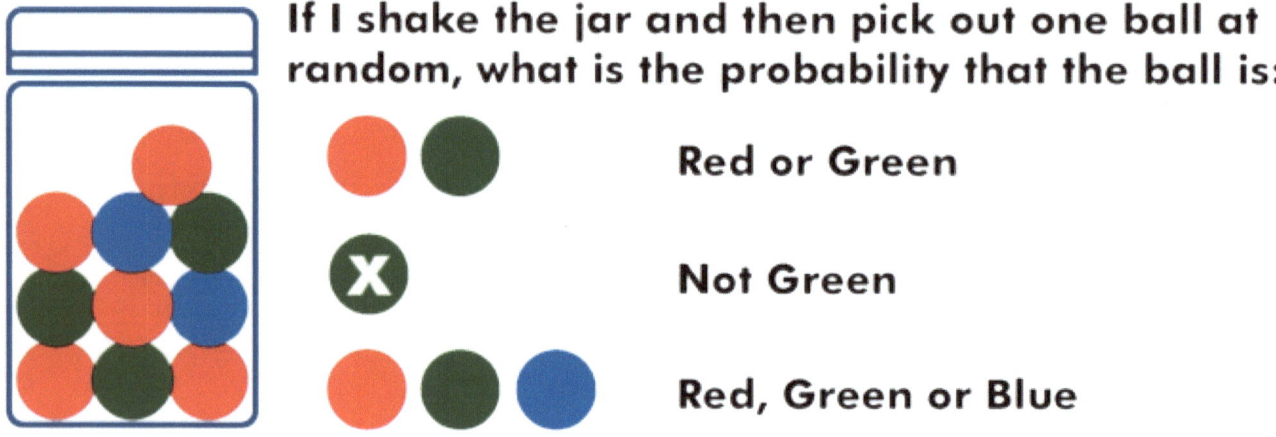

If I shake the jar and then pick out one ball at random, what is the probability that the ball is:

Red or Green

Not Green

Red, Green or Blue

Flip a Coin

If I flip a coin twice, what is the probability of getting heads twice?

A Tree Diagram

The sample space for flipping a coins twice is: {HT, HH, TH, TT).

The probability of getting two heads is $\frac{1}{4}$.

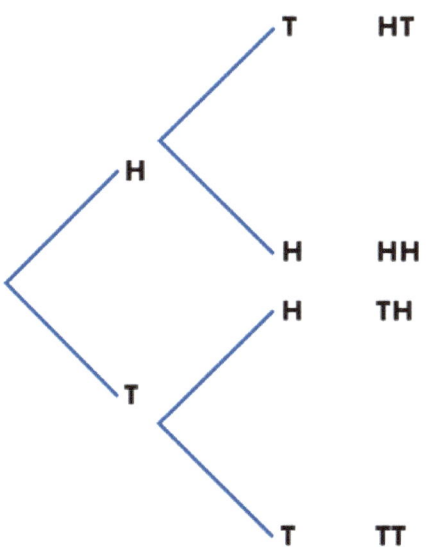

T   HT

What is the probability of flipping two tails?

H   HH

H   TH

What is the probability of flipping a head and a tail?*

T   TT

Sample space for flipping
a coin twice: {HT, HH, TH, TT}

*Any order

Complete the tree diagram for flipping a coin three times.

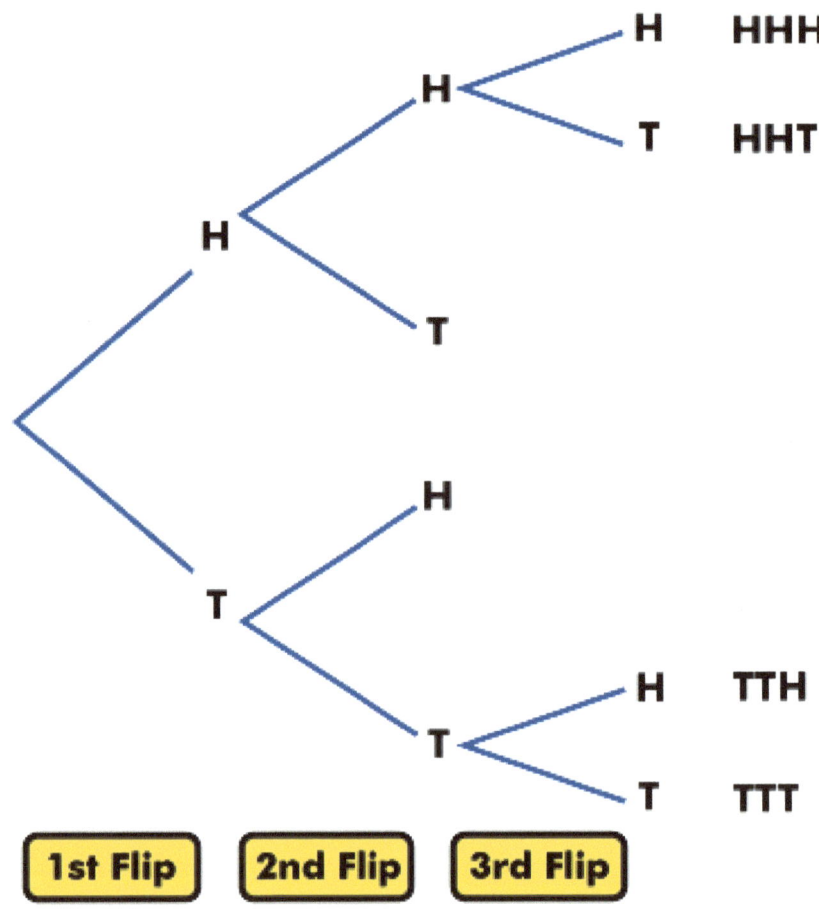

What is the sample space?

Analyzing the Data from a Tree Diagram

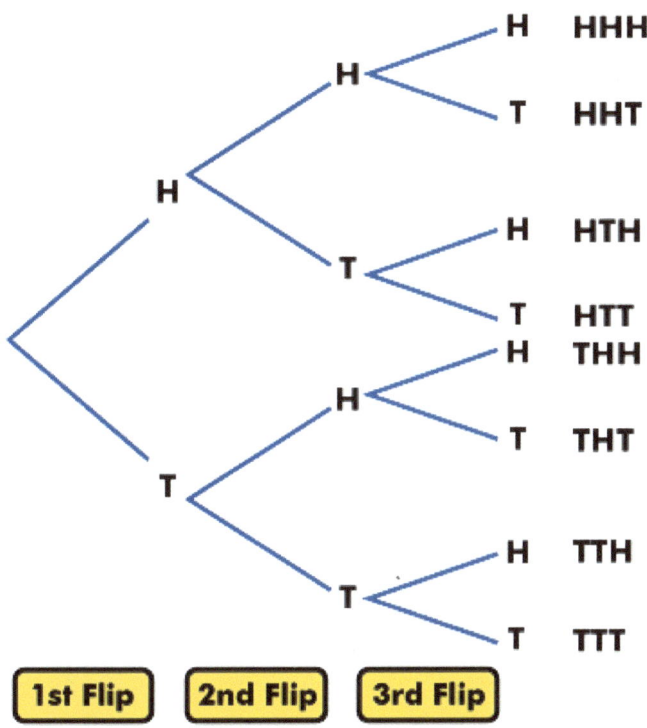

What are the total number of outcomes?

What is the probability of flipping (any order):

1 **Three heads?**

2 **Two heads and a tail?**

2 **At least two tails?**

1st Flip   2nd Flip   3rd Flip

Sample space for flipping three coins:
{HHH, HHT, HTH, HTT, THH, THT, TTH, TTT}

Name_____

Probability Quiz

1 True or false? If an event has a probability of 0, it is impossible. A probability of 1 means certain.

2 What is the probability of choosing a green ball?

A 30%

B 3%

C 20%

D 70%

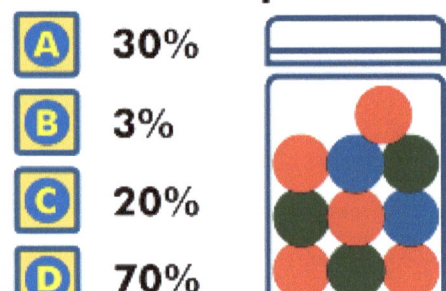

3 You flip a coin twice. What is the probability of flipping two heads?

4 You flip a coin twice. What is the probability of *not* flipping two tails?

Give answers to questions 3 and 4 in decimal form.

Newburyport, MA 01950

1-800-596-3175

OnBoard Academics employs teachers to make lessons for teachers! We create and publish a wide range of aligned lessons in math, science and ELA for use on most EdTech devices including whiteboard, tablets, computers and pdfs for printing.

All of our lessons are aligned to the common core, the Next Generation Science Standards and all state standards.

If you like our products please visit our website for information on individual lessons, teachers licenses, building licenses, district licenses and subscriptions.

Thank you for using OnBoard Academic products.

www.ingramcontent.com/pod-product-compliance
Lightning Source LLC
Chambersburg PA
CBHW050409180526
45159CB00005B/2211